Fleure The Magical Metal Horn U

With

Fluorine Magic

AF871626

Element 9 – Magical Elementals Chemistry Fantasy

By Sybrina Durant

Fleure The Magical Metal Horn Unicorn With Fluorine Magic

Element 9 – Magical Elementals Chemistry Fantasy

Copyrighted 2026

Paperback ISBN 978-1-942740-78-0

BISAC Codes:
JUV036000 - JUVENILE FICTION / Science & Technology
JUV037000 - JUVENILE FICTION / Science Fiction, Fantasy, Magic
JUV002270 JUVENILE FICTION / Fantasy / Dragons, Unicorns & Mythical Creatures

All rights reserved by Sybrina Publishing and Distribution Company.

League City, Texas, United States of America

This book contains material protected under International and Federal Copyright Laws and Treaties. Any unauthorized reprint use of this material is prohibited.

No part of this book may be reproduced or transmitted in any form or by any means, electronic or mechanical, including photocopying, recording, or by any information storage and retrieval system without written permission from Sybrina Publishing and Distribution Company.

Contact Sybrina@sybrina.com.

No Metal
Fleure
The Fluorine-Horned
Unicorn
No Magic

Chapter One: The Chartreuse Unicorn of Verdanmoor

The meadows of Verdanmoor stretched endlessly beneath a sky the color of polished copper, where the moon hung low on the horizon even at midday, casting silver shadows across fields of phosphorescent clover. Ancient aqueducts of copper-threaded obsidian carried mineral-rich water from the Crystalline Heights down into the valley, testament to the old alchemist engineers who had built MarBryn's first civilizations on the philosophy that defined every stone, every spark, every breath of this world: "No Metal, No Magic."

Reclining beside a humming aqueduct pillar, a small but remarkable figure rested in the tall grass. She was impossible to miss. Her compact body radiated a vivid chartreuse yellow-green, almost luminous against the emerald clover, as though she had been assembled from concentrated sunlight filtered through a forest canopy. A pair of magnificent butterfly wings folded loosely at her back, their azure and cobalt panels catching the copper light and scattering it in iridescent fragments across the meadow around her. Her mane cascaded in thick lime and emerald ringlets, adorned with a single teal ornament that clinked softly when she moved. Most striking of all was the horn rising from her brow — sleek, metallic, spiraling to a needle-sharp point that gleamed with a pale metallic lime sheen, ancient and noble and quietly dangerous.

This was Fluere, Elemental of the Ninth Element, unicorn of Fluorine magic, and she was, at this particular moment, deeply annoyed.

"Caspin," she said, her large dark eyes fixed on the wiry young man sitting cross-legged beside her, tinkering furiously with a brass-and-ceramic device the size of a melon, "that device has been beeping for forty-seven minutes."

Caspin Wrale grinned without looking up. His brown fingers were nimble and perpetually ink-stained, his copper-rimmed goggles pushed up onto a forehead that bore a faint scar from a previous experiment gone spectacularly wrong. He was seventeen, human, possessing no magic whatsoever, but what he lacked in elemental power he more than compensated for in chemical engineering genius. The device in his hands was a spectral resonance scanner — his own invention — capable of detecting unusual elemental signatures in the surrounding atmosphere.

"Forty-seven minutes," he confirmed cheerfully, "because for forty-seven minutes there has been an absolutely unprecedented fluorine concentration spike coming from the direction of the Ashveil Plateau. Which is fascinating, because that plateau has been elementally dead for three hundred years."

Fluere lifted her head, her blush-pink-lined ears rotating forward. Her expression shifted from mild irritation to sharp attention. "Show me."

He turned the scanner toward her. The display — a thin panel of luminous selenite crystal embedded in the brass casing, one of MarBryn's earliest display technologies — showed a cascading readout of elemental data. The fluorine signature was impossible to argue with. It pulsed in harsh, spiking waves.

"That is not natural accumulation," Fluere said quietly. She rose to her four small hooves, wings spreading instinctively, their cobalt veins catching the light. "Something is generating it."

"Or someone," Caspin said, snapping the scanner shut and standing. He was a full head taller than her back, though she could look him directly in the eyes when she chose to hover. "There is something else." His grin faded slightly. "There's a secondary signature layered underneath. Corrosive. Organized. I've only ever read about signatures like that in the old Alchemical War texts." He hesitated. "It matches Scorivex."

The name landed in the meadow like a cold stone dropped into still water.

Fluere was silent for a long moment. Her horn glowed faintly at the tip — pale and cold, like starlight — and the temperature around her dropped several perceptible degrees as her hypothermic aura responded instinctively to the spike of emotion she refused to show on her face.

"Scorivex the Corruptor," she said at last, her voice carefully level, "has been imprisoned in the Nullvault beneath the Ashveil Plateau for three centuries."

"Yes," said Caspin.

"And now there is an unprecedented elemental surge coming from the Ashveil Plateau."

"Also yes."

Fluere spread her wings fully. The span was breathtaking — nearly double the width of her body, brilliant azure panels trembling with contained energy. She looked toward the distant dark smudge of the plateau on the horizon, and something ancient and protective and fierce moved behind those warm brown eyes.

"Pack your scanner," she said. "We leave immediately."

Caspin was already throwing his satchel over his shoulder. "I packed twenty minutes ago," he said. "I was just waiting for you to stop napping."

"I was thinking," Fluere said with great dignity, and leaped into the copper sky.

Chapter Two: The Ashveil Plateau and the Stolen Catalyst

The Ashveil Plateau rose from the lowlands like a broken tooth, its ancient basalt flanks streaked with mineral veins of green malachite and silver galena — the bones of the earth made visible, proof that even in this dead geological giant, the elements endured. The plateau had once been a center of alchemical research, three centuries ago, when MarBryn's greatest chemist-sorcerers had attempted to harness Reactive Element Convergence — the forced fusion of volatile elemental magic to generate unlimited energy. The experiment had created Scorivex instead: a being of pure corrosive chemistry, born from a catastrophic reaction between seventeen elemental streams simultaneously. It had taken the combined power of four Elemental unicorns to contain him in the Nullvault, a prison of inert noble-gas-infused stone designed to suppress all elemental activity.

Fluere banked on a thermal current above the plateau's edge, Caspin secured in a harness she had conjured from compressed ice and hardened fluoride crystal — one of her more practical frost-conjuring applications that she had developed specifically for traveling with her non-flying companion. Below, the plateau's surface was wrong. Dark cracks split the basalt in geometric patterns too regular to be natural, and from those cracks rose thin, hissing columns of pale gas.

"Hydrofluoric vapors," Fluere said grimly, analyzing the chemical signature with the sensitivity her elemental nature provided. "Someone has been running acid reactions into the Nullvault's foundation. Systematically. For months, possibly."

"Dissolving the noble-gas matrix from below," Caspin said, his scanner working furiously. "Clever and horrible. The matrix is what keeps Scorivex suppressed. Without it—"

A sound rolled up from the plateau. Low, grinding, like tectonic plates shifting — except it was not geological. It was intentional. It was deliberate.

And then, from a crack wide enough to swallow a building, the first of Scorivex's Acid Drones emerged.

They were constructs — ancient alchemical automatons that Scorivex had apparently been assembling within his prison during his long confinement, building them from dissolved mineral deposits and his own corrosive essence. They stood roughly twice the height of a human, bipedal, their bodies a patchwork of fused mineral and hardened acid residue, their hollow eyes glowing with acrid yellow-orange light.

"How many?" Fluere asked.

"Scanner says nine," Caspin said. "Also, I would like to lodge a formal complaint about our morning so far."

Three Drones spotted them in the air and launched streams of caustic liquid upward with launcher-arms built into their mineral torsos.

Fluere acted. Her horn blazed with cold white light, and she swept it in a broad arc. A wall of dense ice manifested thirty feet above the plateau, angled perfectly to deflect the caustic streams away into the basalt. The ice hissed and pitted where the acid struck, but held.

"Landing," she announced.

"That is the opposite of what I would choose!" Caspin called as she folded her wings and dropped.

She hit the plateau with controlled force, wings snapping open at the last moment to bleed velocity. Her forehooves — split and dexterous, closing on the basalt with practiced precision — touched down as two Drones charged. She swung her horn toward the ground between them and a jagged frost-spike ridge erupted from the plateau surface, catching the first Drone's legs and locking them in crystalline ice. The second reached her. She stood her ground, flooded her hypothermic aura to maximum intensity, and the Drone's mineral joints seized as temperatures around her plummeted to near-absolute in a six-foot radius. It toppled, rigid.

Caspin landed, hit a roll on the basalt, came up with a device in each hand — a compressed alchemical disruptor of his own design that fired bursts of neutralizing base solution, devastating against acid-based constructs. He fired three times in quick succession, each burst dissolving a Drone's core housing.

Five minutes later, nine Drones lay destroyed across the plateau.

At the center, the Nullvault's great seal — a disc of pale grey stone twenty feet across, carved with the original containment formulae in precise elemental notation — had a crack running diagonally across it. Along the crack, freshly placed and still faintly steaming, was a Reactive Catalyst — a dense crystalline rod of engineered fluoride compound that had been used to initiate and sustain the acid dissolution process.

Fluere lifted it carefully in her split forehooves, examining it. The craftsmanship was extraordinary. This was not ancient technology.

"This was made recently," she said. "By someone who understands fluorine chemistry at a very advanced level." She met Caspin's eyes. "Scorivex did not make this from inside the vault. He had help."

Caspin looked at the formulae etched into the catalyst rod and his face went pale. "Fluere," he said quietly, "I recognize this notation style. It is from the Chemarcane Academy."

Chapter Three: The River Road to Luminance City

The River Veloris ran like a ribbon of liquid mercury through the lowlands south of Ashveil, its mineral-rich waters supporting floating communities built on buoyant ferrowood platforms — half timber, half structural iron, entirely characteristic of MarBryn's hybrid ancient-and-modern philosophy. The communities were called Driftsteads, and they had been trading along the Veloris since before the Alchemical Wars.

Fluere and Caspin traveled the river road toward Luminance City, one of MarBryn's greatest centers of knowledge, where the Chemarcane Academy rose above the city's crystalline skyline like a monument to everything the world knew and still sought to understand. They moved at a ground pace, Fluere's wings folded, the better to travel inconspicuously along a road busy with merchants hauling mineral goods in carbon-fiber carts beside farmers with copper-shod mules.

"Tell me about the Chemarcane Academy," Fluere said, watching a Driftstead float slowly past on the river below the road embankment, lanterns already lit in the early afternoon dimness under the plateau's shadow.

Caspin, walking beside her, tugged his goggles off his forehead and polished them absently. "Founded eight hundred years ago. Trains chemist-sorcerers — people who blend elemental magic with scientific chemistry. The notation on that catalyst is Second-Era Chemarcane academic script, which means whoever made it was either trained there or stole the methodology from someone who was." He paused. "The Academy's current Arch-Chancellor is a man named DrovenSarric. He has been in that role for eleven years and in that time he has made the Academy increasingly... closed. Secretive."

"You know a great deal about it," Fluere observed.

Caspin was quiet for a moment. "My mother was a faculty member there," he said. "Before she was dismissed. DrovenSarric dismissed her for what he called dangerous theoretical work. She was researching containment breach scenarios for the Nullvault." A pause. "She was right to be concerned, apparently."

Fluere turned her warm dark eyes on him. "I am sorry," she said simply, and meant it.

Their road curved around a limestone bluff, and there, sitting on a carved rock at the roadside with her legs dangling and a battered tome open across her knees, was a young woman they nearly walked past entirely. She had close-cropped silver hair despite being perhaps twenty, tawny skin, and eyes of an unusual violet hue that immediately identified her as a Sparkmage — those born with electrical and light manipulation abilities. She wore the dusty traveling clothes of someone who had been moving for days and looked up at their approach with an expression of equal parts exhaustion and relief.

"Fluere?" she said. "Fluere of Verdanmoor? The Fluorine Elemental?"

Fluere stopped. "I am she. And you are?"

"Valeria Vorn." The woman closed the tome and stood, extending a hand. "Chemarcane Academy — former student. I left six months ago when I discovered that Arch-Chancellor Sarric was running unauthorized elemental experiments in the Academy's deep vaults." She held up the tome. "I took evidence. I have been trying to find someone who would listen to me ever since. And then I heard an elemental signature burst from the Ashveil direction this morning and thought — if anyone is moving toward this problem, it would be someone like you."

Caspin looked at Fluere. Fluere looked at Caspin.

"Walk with us," Fluere said.

Valeria fell into step beside them, tucking the tome under her arm. "How bad is it at Ashveil?"

"The Nullvault seal is cracked," Caspin said. "Someone planted a fluoride catalyst to accelerate the dissolution of the containment matrix."

Valeria stopped walking. "That means—"

"Scorivex is not fully free yet," Fluere said. "But the process is well underway. We need to reach the Academy, determine the extent of Sarric's involvement, and find a way to re-establish the containment before the matrix fails entirely." She paused. "How long do we have?"

Valeria opened the tome, scanning a page of dense notation. "Based on what I know of the matrix's composition and the acid exposure rate you described?" She looked up grimly. "Eleven days. Perhaps twelve."

Chapter Four: Luminance City and the Vaulted Academy

Luminance City deserved its name. It rose from the confluence of three mineral rivers in a cascade of crystalline architecture — buildings of selenite, quartz-composite, and carbon-lattice glass that caught the twin-moon light and scattered it in permanent prismatic glory across the streets below. The city's infrastructure was a masterwork of applied chemistry: filtration towers of layered zeolite stone cleaned the air, copper-catalytic street lighting burned without flame, and beneath every street ran the ancient Vein-Lines — tubes of conducting mineral through which elemental communication signals traveled across the entire city. The philosophy "No Metal, No Magic" was perhaps most visible here, where every building, every system, every comfort was an expression of elemental chemistry made practical.

Fluere was drawing stares. A chartreuse unicorn with butterfly wings was not an everyday sight even in Luminance City, and she was aware of it with quiet, unhurried dignity — the teal ornament in her mane catching the crystal-scattered light, her lime-colored metallic horn gleaming. She kept her wings folded tight and walked between Caspin and Valeria through the market district, which smelled of mineral dust, chemistry-reagent shops, and fresh river bread.

"We need access to the Academy's public archives first," Valeria said, steering them down a side street. "Specifically the theoretical vault logs from the last three years. If Sarric has been siphoning containment knowledge, the logging system will show it — he cannot erase logs from the public archive terminals without leaving traces. I know that because I designed part of that logging system as a third-year student project."

"You designed Academy infrastructure as a student project?" Caspin said, impressed.

"It was extra credit," Valeria said, slightly defensive.

The public archive was housed in the Academy's outer wing — accessible without special clearance, a vast reading room of polished malachite and selenite where scholars and students worked at terminals of mineral-crystal computing, a technology that had emerged from understanding the electrical properties of crystalline mineral structures. The Academy itself loomed above them through the archive's tall windows: seven towers of increasing height, the tallest housing the Arch-Chancellor's offices at its apex, its spire catching both moons' light simultaneously and throwing twin silver beams across the city.

Valeria took a terminal and worked quickly, her violet eyes scanning data at remarkable speed. Caspin set up beside her, interfacing his scanner with the terminal to pull deeper data layers. Fluere stood behind them both, reading over their heads, one split forehoof resting on Caspin's chair back.

"Here," Valeria said sharply. "Deep access requests to the Nullvault containment theoretical archive. Seventeen access events over three years. All logged under a research clearance code that was officially decommissioned four years ago."

"A ghost clearance," Caspin said. "Someone inside the Academy maintained access privileges from a cancelled account." He paused. "That requires either the Arch-Chancellor's direct authorization or a very sophisticated forgery."

"Neither makes Sarric look good," Valeria said.

"It does not," said a new voice from behind them.

They turned. Leaning against the archive doorway with the casual confidence of someone who owned any room they entered was a tall figure — an older man, perhaps fifty-five, with the muscular build and calloused hands of a field alchemist, his dark skin weathered and his grey-streaked hair braided back with copper wire. He wore a long coat of treated ferrowoodweave, and at his belt hung a brace of alchemical grenades — small globes of compressed reactive reagent — beside a long staff of iron-composite topped with a selenite crystal.

"My name is Gordas Veln," he said. "I was Arch-Chancellor of this Academy before DrovenSarric replaced me eleven years ago." His eyes moved to Fluere with something like relief. "I have been investigating Sarric's activities for three years. I heard an elemental surge from Ashveil this morning, and I came here because I suspected someone would." He pushed off the doorway. "Whatever you have found, we need to move quickly. Sarric knows his catalyst was activated. By now he also knows it was found."

As if to confirm this, the Academy's warning tone — a resonant hum carried through the Vein-Lines — began sounding throughout the building. Across the archive, scholars looked up in confusion.

"Academy lockdown protocol," Valeria said, her violet eyes hardening. "Sarric is sealing the building."

"Then we go through it, not around it," Gordas said, unshouldering his staff.

Fluere stepped forward, her horn tip igniting with cold white light. Outside the archive windows, the temperature had already begun to drop.

"Lead the way, former Arch-Chancellor," she said.

Chapter Five: The Deep Vault Battle

The Academy's lower levels descended through the limestone bedrock like a vertical timeline of MarBryn's scientific history — the higher floors contemporary, crystalline, full of living research; the lower floors progressively older, their walls shifting from selenite composite to raw stone carved with elemental notation in ancient scripts, torch-lit by chemical-flame brackets that burned without fuel using century-old catalytic reactions. It was beautiful and it was eerie in equal measure.

Sarric's first wave of defenders met them at the third sub-level.

They were not constructs like the Acid Drones. They were human — Academy security personnel, armed with alchemical-discharge weapons: staffs that fired concentrated bursts of reactive chemistry, capable of burning through ferrowood or mineral composite. A dozen of them formed a line across the corridor.

"Stand down," their captain called. "By order of Arch-Chancellor Sarric, this facility is under restricted access."

"By order of the actual laws of MarBryn," Gordas returned calmly, "DrovenSarric has no authority to obstruct investigation of a containment breach at a classified Nullvault site. Stand your people down."

The captain raised his staff.

What followed was fast, loud, and extraordinary.

Valeria moved first — both hands raised, violet light crackling between her fingers as she was a Sparkmage in full form. She discharged a controlled electromagnetic pulse that knocked six staffs offline simultaneously, their crystal capacitors blown. Gordas was already moving, his iron-composite staff spinning in his hands with the practiced ease of decades, disarming two guards with efficient strikes.

Caspin threw one of his neutralizing solution grenades — a design he had upgraded from the Acid Drone encounter — directly into the corridor ceiling above the guards. It burst in a wide spray, not harmful to humans but coating every alchemical discharge weapon within range in base-neutralizing compound, rendering them inert. "Chemistry problem, chemistry solution," he said, with moderate satisfaction.

The guards directly ahead of Fluere charged.

She spread her wings in the corridor — not to fly, but to create a sudden wall of presence and the blast of displaced air from powerful strokes. The temperature in the corridor dropped twenty degrees in three seconds as her hypothermic aura expanded to its full range. The lead guard slowed visibly, his movements becoming sluggish. She aimed her horn with surgical precision and fired three focused beams of cold: the first froze the captain's weapon-hand to his staff grip, the second and third created ice-bridges across both guards' legs, locking them to the floor. Non-lethal, contained, efficient.

In forty seconds the corridor was clear.

"That," Caspin said to Valeria, slightly out of breath, "is what she looks like when she actually tries."

The deep vaults opened below the fifth sub-level in a cathedral-scale chamber of ancient stone, its ceiling lost in darkness above chemical-flame pillars. This was where Scorivex had originally been studied, where the Academy's founders had catalogued the knowledge that led to the Nullvault containment. The chamber's walls were covered floor-to-ceiling in original elemental notation — a visual record of the Alchemical Wars' chemical discoveries, both brilliant and terrible.

And at the chamber's center, surrounded by three years' worth of unauthorized experimental apparatus, stood DrovenSarric.

He was a thin man, sharp-featured, wearing Arch-Chancellor's formal robes that seemed incongruous against the raw-stone violence of the chamber around him. His eyes, when he looked up, held the particular brightness of absolute conviction.

"Fluere," he said, as though expecting her. "The Fluorine Elemental. Of course."

"Sarric," Gordas said, his voice tight with controlled anger. "What have you done?"

"What our founders should have done three hundred years ago," Sarric said calmly. "Scorivex is not an enemy. He is the greatest naturally occurring chemical intelligence in MarBryn's history. A mind built from reactive chemistry, from the interaction of seventeen elemental streams. He should not be imprisoned. He should be studied."

"He destroyed four cities," Valeria said.

"Because he was frightened and contained. A reaction naturally seeks equilibrium. I am helping him reach it."

Fluere's horn glowed. "You are releasing a force that will achieve that equilibrium by dissolving everything around it," she said. "Including the people."

Sarric's hand moved to a device on the table beside him — a secondary catalyst, larger than the one at Ashveil. His finger was on its activation trigger.

"The vault seal is already seventy percent dissolved," he said. "In approximately eight days, Scorivex will be free regardless. I am asking you to let it happen peacefully. Or—" His finger pressed the trigger halfway. "I finish the process in the next thirty seconds."

Fluere did not hesitate. She acted on fluorine instinct — molecular manipulation, the ability she rarely used because it required deep focus and tremendous precision. She targeted not Sarric but the catalyst device itself: restructured the molecular bonds of its reactive core from the outside in, turning the crystalline fluoride compound inert in three precise, focused pulses from her horn. The device's activation light went dark.

Sarric stared at it.

Gordas crossed the chamber in four strides and had Sarric's arms pinned before the man recovered from his surprise. "It is over, Droven."

"It is not over," Sarric said, very quietly. "The vault seal is still dissolving. You cannot stop that without a full recasting of the noble-gas matrix. Do you even know how to do that?"

Fluere looked at Gordas. Gordas looked at her. And both looked at the notation on the walls — the original containment formulae, written in the founders' hands.

"No," Fluere admitted. "Not yet. But we are going to learn."

Chapter Six: The Formulist of Deepmere

With Sarric in Academy custody and Gordas re-establishing himself as acting Arch-Chancellor to the visible relief of the majority of Academy faculty, the team had access to the full Academy resources. What they quickly established was that re-casting the noble-gas containment matrix required not only the formulae from the deep vault walls but a living expertise in ancient containment chemistry that had been largely lost.

"There is one person," said a senior Academy researcher, a small, fussy woman named Dorveth who wore three pairs of spectacles simultaneously and smelled permanently of sulfur, "who might be able to reconstruct the methodology. A woman named SiaConnwick. She calls herself a Formulist — essentially an archaeo-chemist who specializes in reconstructing ancient chemical-magical processes. She lives in Deepmere, in the Karst Caverns."

"Of course she does," Caspin said.

Deepmere was a full day's flight south, through the mountain range the maps labeled as the Orewall Peaks — a chain of iron-rich mountains whose exposed mineral veins gave them a rusty, almost blood-red coloring against the sky. The region was known for extreme elemental weather: magnetic storms, mineral precipitation, and the occasional spontaneous chemical-fire bloom from exposed pyrite seams.

Fluere flew the route with Valeria beside her — the Sparkmage had a degree of personal flight through electromagnetic lift, rising on her own crackling violet field — while Caspin and Gordas traveled by the mountain express route via one of MarBryn'sVeloris Rail lines, a rail system using magnetic-mineral repulsion to run at considerable speed through mountain tunnels. They agreed to converge at Deepmere's northern cavern entrance.

The flight through the Orewall Peaks was breathtaking and treacherous. Magnetic storms rolled through the peaks unpredictably, visible as crackling curtains of blue-white energy between the red-iron summits. Valeria navigated them with practiced electromagnetic sensitivity. Fluere simply flew through, her hypothermic aura creating a bubble of cold that disrupted the magnetic patterns immediately around her.

"You know," Valeria called across a gap between two peaks, her voice raised above the wind, "I have never flown alongside a unicorn before."

"I would imagine not," Fluere called back, banking on a mountain thermal. Her wings — cobalt and azure, dots of white marking them like scattered stars — caught the updraft with the grace of something that had been flying since before it could walk. "What made you leave the Academy when you discovered what Sarric was doing? You could have stayed silent. Many did."

Valeria was quiet for a moment, riding her electromagnetic lift. "My mother always said that the greatest danger in a world built on the philosophy of No Metal No Magic is forgetting what the second half means. The magic matters as much as the metal. The why matters as much as the what." She looked across at Fluere. "Scorivex being free would mean no civilization. That matters."

They found the Deepmere caverns exactly where the researcher's directions indicated: a honeycomb of karst passages opening in the southern face of the final Orewall peak, their entrances marked with chemi-luminescent mineral paint in colors that indicated safe passages from unsafe ones — another layer of applied chemistry, here in a place where someone had decided that technology and wilderness could share a home.

SiaConnwick met them at the cavern mouth. She was younger than the title Formulist implied: perhaps thirty, with deep tanned skin, natural hair wrapped in protective cloth, and hands that were permanently stained with indicator compounds in three different colors. She wore a coat covered in pockets, all full. She looked at Fluere with the specific expression of someone encountering the confirmation of a long-held theory.

"A Fluorine Elemental," she said. "Fascinating. Come in. I already know why you are here — I heard the Ashveil signature yesterday and I have been running the math since." She turned and disappeared into the luminescent cavern. "The matrix recasting is possible but it requires a living fluorine source to anchor the noble-gas chemistry. Specifically," she called back over her shoulder, "it requires you, Fluere."

Chapter Seven: What the Old Formulae Demanded

The interior of Sia's cavern complex was the most extraordinary intersection of ancient and modern technology Fluere had seen in one place. The cave walls were the original stone, rich with mineral deposits and ancient chemical notation — some of it identical to the Academy deep vault's writing. Nestled among the stalactites and mineral formations were modern instruments: spectral analyzers, crystal-display computing terminals, a small but sophisticated chemical laboratory with equipment that had been brought in piece by piece and carefully installed. Fossil samples sat beside contemporary synthesis chambers. It was, Fluere thought, exactly what MarBryn's philosophy looked like in its most honest form.

Gordas and Caspin arrived through the north entrance two hours after Fluere and Valeria, Caspin with new burn marks on his coat from a mineral-fire bloom the Rail had passed through and Gordas with the slightly vindicated expression of someone who had suspected train travel would be problematic.

The five of them gathered around Sia's central work table as she laid out her analysis.

"The noble-gas containment matrix works like this," she said, pressing a display panel that activated a three-dimensional crystalline model above the table — an Academy-tech projector she had apparently modified with her own improvements. "Krypton and xenon compound layers form the primary suppression envelope. These compounds are chemically inert by nature. No reaction means no pathway for Scorivex's corrosive chemistry to operate through. He cannot dissolve what will not react." She manipulated the model. "The acid attacks from below degraded those layers by introducing fluoride reactants — specifically attacking the xenon compound bonds."

"Using my element against the containment," Fluere said.

"Yes. Which is also why restoring it requires your element in a controlled, constructive application." Sia looked directly at Fluere. "The original formulae require a Fluorine Elemental to synthesize a re-bonding compound — essentially, a fluoride-xenon restorative chemistry that repairs the matrix bonds rather than dissolving them. It is the same chemistry, used oppositely." She paused. "It requires you to channel your molecular manipulation ability for an extended period — perhaps two hours of continuous working. And it will need to be done at the Nullvault site directly."

"What is the risk to Fluere?" Gordas asked.

Sia's pause was brief but noticeable. "Sustained molecular manipulation at that scale is physically demanding. She will be exhausted after. Genuinely, deeply exhausted." Another pause. "And she will need to do it while standing over an active breach point. Scorivex, at seventy-plus percent dissolution, will be able to sense and target her from below the vault. He will fight the recasting."

"Then we defend her while she works," Caspin said, without hesitation. He looked around the table. The same expression was on every face.

Valeria cracked her knuckles, violet light sparking between her fingers. "Obviously."

"There is one additional requirement," Sia said, almost reluctantly. "The recasting requires a full noble-gas signature set — not just fluoride-xenon chemistry. It needs an Elemental signature anchor from the fluorine side and a stabilizing signal from the krypton-side chemistry. Krypton is Element 36." She looked at Gordas. "I know what you are, former Arch-Chancellor. Everyone in archaeo-chemistry knows. You have been hiding it for eleven years but your field research signatures are unmistakable."

Gordas was very still.

"Gordas," Fluere said gently, "what is she saying?"

The older man closed his eyes briefly and then straightened, copper-wire braids shifting against his coat. "I am a Krypton Elemental," he said quietly. "I hid it from Sarric when he took the

Academy, knowing that he would have used me for his Scorivex projects. I am not powerful in the way you are, Fluere. I do not fight with it. But for anchoring chemistry — for stabilization — yes. It is what I am."

Fluere reached out and placed one split forehoof gently on his arm. "Good," she said. "Then we can do this."

Sia closed the display. "You have approximately six days. I will travel with you to Ashveil. I will guide the chemistry from outside the breach point while you two work inside it." She was already putting on her coat and loading her pockets with sample vials and instruments.

"You said five people could come," Caspin said to no one in particular.

"I did not say that," Fluere said.

Chapter Eight: The Road Back to Ashveil

The journey back to Ashveil Plateau took them two days, traveling as a group of five now — Fluere, Caspin, Valeria, Gordas, and Sia — moving via a combination of flight, the Veloris Rail northern line, and a hired mineral-cart for the final stretch of road where the rail ended and the plateau's proximity began disrupting the magnetic repulsion system that powered the trains.

The disruption was new. When Fluere had traveled this road four days prior, the rail had functioned normally all the way to the Verdanmoor junction. Now the interference extended thirty miles further south, and it was getting stronger.

"The dissolution is accelerating," Sia said from the back of the mineral-cart, three instruments running simultaneously across her coat-covered knees, her stained fingers tabbing through readouts with practiced speed. "We lost a day when I underestimated the secondary acid-feed reactions. The catalyst Sarric installed at the Ashveil site had a delayed-release secondary chamber. It has been running since we removed the primary rod."

"Can it be stopped remotely?" Caspin asked. He was beside her, his own scanner open, cross-referencing her data.

"Not without triggering an immediate full breach," Sia said. "It was designed that way deliberately. Tamper-resistant. Sarric was thorough." She said the last word with the particular bitterness of someone who respected competence in an enemy she despised.

"How long do we actually have?" Gordas asked from the cart's front bench, where he sat beside the driver — a weathered Driftstead woman named Orla who had agreed to take them to the plateau base for what Gordas described as a generous fee and what Orla described as nowhere near enough.

"Four days," Sia said. "Perhaps three and a half."

The cart was quiet for a moment.

"We had eleven days four days ago," Valeria said.

"Yes," Sia agreed. "The math is unpleasant."

Fluere was flying alongside the cart rather than riding in it, her wings keeping her at a gentle glide-pace matched to the cart's speed, her mane ringlets streaming back in the wind of their passage. She had been quiet since Deepmere, processing everything Sia had told her about the recasting requirements. Two hours of continuous molecular manipulation. Standing over an active breach point. Scorivex fighting her the entire time.

She was not afraid — not exactly. But she was aware, in the way that quietly confident creatures sometimes became aware, of the specific shape of what was being asked of her. The chartreuse light of her coat seemed somehow more vivid against the darkening sky as evening came on, as though her body was already preparing for the expenditure.

Caspin appeared at the cart's side rail, leaning out to speak to her. "You have been flying without talking for three hours," he said. "That is not like you."

She glanced at him sideways, those large dark eyes warm. "I am thinking."

"You said that last time and you were napping."

"I was thinking and napping," she said, and a small smile curved her mouth — the first in several hours. "Caspin. The neutralizing solution grenades you have. How many do you have left?"

He calculated. "Fourteen standard, six enhanced concentration. Why?"

"Scorivex's Acid Drones at Ashveil — the nine we defeated were a scouting force. The dissolution is now far more advanced. When we arrive, the breach will be significantly wider and he will have had four additional days of partial freedom to create more constructs." She looked toward the dark smear of the plateau on the horizon. "We are going to face a much larger defensive force before I can even reach the vault point to begin the recasting."

"I know," Caspin said simply.

"It will be dangerous."

"Also know that."

"Caspin." She stopped gliding and hovered, the cart pulling slightly ahead for a moment before Orla slowed without being asked — the Driftstead woman apparently having good instincts. Fluere looked at her companion directly. "You have no magic. You have no elemental power. You have chemistry knowledge and devices and a great deal of courage, and you have been at my side for two years without a single moment of hesitation. I want you to know that I see that. That I am grateful for it."

Caspin looked at her for a long moment. His ink-stained fingers tightened on the rail. "My mother always said that the most important thing a person could be was useful to something good," he said. "I think I finally understand what she meant." He straightened. "I am going to build something tonight while we camp. Something that will help with the Drones."

"What?"

He grinned, and it was the particular grin of a chemical engineer with an idea. "You will see."

They camped at the base of the plateau's southern approach, where a formation of wind-carved basalt columns provided shelter and cover. Sia spent the night running final formulae calculations by chemi-luminescent lamp, her lips moving silently over ancient notation. Gordas sat apart from the group and meditated in the still way of someone accessing their elemental nature, preparing. Valeria maintained a perimeter, her violet sparkmage senses extended outward into the dark, watching for Drone scouts.

Fluere slept — genuinely this time — curled in the shelter of the largest basalt column, wings folded over herself like a self-made blanket. She dreamed of open meadows and twin-moon skies and the smell of phosphorescent clover, and it steadied something deep inside her.

In the morning, Caspin showed them what he had built: a Broadfield Neutralizer — a backpack-mounted device with an omnidirectional emitter that could dispense neutralizing base-solution in a thirty-foot radius burst rather than a targeted throw. Improvised from his remaining supplies and components stripped from non-essential parts of his scanner.

"One use only," he warned. "But it will clear a large cluster of Drones in a single activation."

"You built that overnight," Sia said, with the specific expression of one scientist encountering another scientist's unexpected competence.

"I had a good idea and the right materials," Caspin said, strapping it on. "Also I did not sleep at all, so everyone please be impressed today so that it was worth it."

They looked up at the Ashveil Plateau in the grey pre-dawn light. From its cracked surface, at least twenty columns of pale hydrofluoric vapor were rising now, and the sound that had been a grinding rumble four days ago had deepened into something that resonated in the chest — a subsonic pressure that made the basalt under their feet tremble intermittently.

From the largest crack in the plateau surface, a slow, dark shape was beginning to emerge. Not a Drone.

Something much larger.

"That," Gordas said, his voice steady, "is Scorivex attempting a partial emergence. He is not fully free — but the breach is wide enough for a projection of his essence."

The shape was vast and wrong, a column of dark corrosive chemistry made semi-visible, crackling with acid-orange light at its edges, its form shifting and reforming constantly as it pushed through the broken seal.

Fluere spread her wings to their full magnificent span, azure and cobalt blazing in the pre-dawn grey, her horn igniting with cold white fire.

"Then we move now," she said. "All of us. Together."

Chapter Nine: The Recasting

What came down from the plateau at them was a wave.

Forty-three Drones — Caspin counted them later, with the specific thoroughness of someone who processes fear through quantification — poured from the cracks in the basalt surface as the group climbed the plateau's access road, a switchback path of ancient stone steps carved by the original alchemist-engineers three centuries ago. The Drones were larger than the first nine, their mineral-composite bodies showing more sophisticated construction. Scorivex had been busy.

Valeria met the first wave at the plateau's edge.

She planted both feet on the ancient stone, raised both arms, and released a Sparkmage's full-power electromagnetic discharge — not the controlled pulses she used in the Academy corridor but a genuine unleashing, violet-white light blazing from her hands in a broad arc that crackled across thirty Drones simultaneously and fused their mineral joints with induced electromagnetic heat. Twenty-three dropped instantly. Seven more staggered, damaged. Valeria dropped to one knee, breathing hard. "That is my limit for twenty minutes," she said tightly.

Gordas moved forward. He was not a combat elemental — had said so clearly — but he carried his iron-composite staff with the authority of someone who had once been asked to defend an Academy and had learned enough to do it. He worked methodically, staff striking and sweeping, while releasing controlled bursts of his krypton elemental ability in a form Fluere had not seen before: a stabilizing inert-gas field that disrupted the acid-chemistry operating Drones' internal systems, causing them to stall and stumble without the combat drama of Valeria's lightning.

"Efficient," Sia noted, moving behind him toward the vault seal at a careful run, her instruments already out.

"I have had eleven years to practice quietly," Gordas said between staff-strikes.

Caspin activated the Broadfield Neutralizer.

The thirty-foot radius burst of concentrated base-solution erupted outward from his position in a billowing white cloud, and every Drone within its reach — twelve of them, converging from the east — dissolved from the outside in with a sound like heavy rain on stone. Caspin exhaled a long, slow breath. "Worth it," he said quietly, and discarded the spent device.

That left eight Drones and the enormous semi-emergent projection of Scorivex, which had now risen forty feet above the vault seal crack, its acid-orange edges flickering as it strained against the remaining fraction of the containment matrix.

Fluere flew directly at the projection.

She did not hit it — she swept beneath it, wings driving her in a tight spiral around the vault seal crack's perimeter, her horn drawing a circle of intense cold in the air at twenty feet altitude. Snowstorm teleportation activated: she merged with the blizzard-fragment she had just created, vanished from the east side of the crack, reappeared instantaneously at the west side, and drew the cold-arc from the opposite direction. Two half-circles of intense cryomancy met and completed a ring around the vault crack — a containment aura of frozen air that crystallized the hydrofluoric vapor columns solid and cut Scorivex's projection off from its feed-source below. The projection shrieked — a sound like metal dissolving in acid — and contracted, diminished but not destroyed, hovering above her in a churning mass of corrosive fury.

"Gordas!" Fluere called. "Now! Come to the seal!"

He broke from the last Drones — Valeria, recovered enough, hit them with precision electromagnetic pulses to clear his path — and crossed the plateau at a run to stand at the vault seal beside Fluere, who had landed at its center. She looked at Sia.

"Talk us through it," Fluere said.

Sia was already at the seal's edge, instruments reading, the ancient recasting formulae from the Deepmere analysis displayed on a portable crystal display propped against the stone. "The formulae requireFluere to initiate molecular restructuring of the exposed matrix bonds starting from the central crack point. Gordas, you begin the krypton-stabilization field simultaneously and maintain it as a constant. The chemistry needs both signatures running in harmony — fluoride reconstruction and noble-gas stabilization. If either of you stops before the matrix is complete, the partial work collapses."

"Two hours?" Fluere confirmed.

"Approximately. I will monitor the matrix integrity and tell you when it is complete. Do not stop until I tell you." Sia looked up at the churning acid projection above them. "We need to keep that from reaching you."

"I have it," Valeria said, arriving at a run. She planted herself ten feet from the seal, violet light building between her hands continuously — not a discharge but a sustained electromagnetic barrier angled upward, a sparkmage's defensive application, converting the acid-orange essence of Scorivex's projection into inert light wherever it tried to descend. "Go. Both of you. I can hold this."

Fluere looked at her. "If it becomes too much—"

"It will not," Valeria said, with the voice of someone making it true through the act of saying it.

Fluere lowered her horn to the vault seal's cracked center. Beside her, Gordas placed both palms flat on the stone and closed his eyes, copper-wire braids falling forward, his krypton field rising around him in a subtle shimmer of inert silver-blue.

And Fluere began.

Molecular manipulation at this scale was not dramatic from the outside. It looked like nothing more than a unicorn standing still, horn touching stone, eyes closed, while the faintest possible cold glow moved like a slow liquid tide outward from the contact point into the crack in the seal. But from inside the working it was everything — a precise, constant, exhausting conversation with the chemistry of the stone itself, finding broken bonds and restoring them, introducing fluoride-xenon chemistry into the matrix structure that had been degraded by acid, rebuilding the noble-gas suppression envelope molecule by molecule while Gordas's krypton field held the framework stable.

Scorivex fought her.

The pressure from below the seal was immense — an acid intelligence three hundred years old throwing everything it had against the restoration. Fluere felt it as physical force, a corrosive resistance pushing back against every bond she repaired. Her hypothermic aura flared and dimmed, flared and dimmed, the cold around her cycling with the rhythm of effort sustained beyond comfort.

Caspin stayed at her right side throughout. He could not help with the chemistry. Instead he did the thing he was best at: he watched, and he measured, and he kept a continuous quiet count of Sia's readings, telling Fluere in a calm, steady voice every fifteen minutes what percentage of the matrix was restored. A number to hold onto. A landmark in the dark.

At ninety minutes, Valeria's barrier flickered.

The acid projection had been systematically testing her defenses for an hour and a half and it found a gap. A tendril of corrosive essence drove through and struck Valeria's left shoulder, burning through her jacket and into skin. She did not drop the barrier. She made a sound that was not quite a word and held, her right arm maintaining the electromagnetic field while her left arm hung useless and smoking.

"Valeria!" Caspin moved toward her.

"Hold," she said through her teeth. "Sia — how long?"

"Twenty-two minutes," Sia said, her voice professionally calm and her hands shaking visibly.

"Then I hold twenty-two minutes," Valeria said.

She held twenty-two minutes.

When Sia said complete, Fluere lifted her horn from the vault seal, and the world swam around her in a wave of profound exhaustion unlike anything she had experienced before. The seal beneath her glowed briefly — an internal silver-blue light as the restored noble-gas matrix lit its own confirmation — and then went dark and solid and utterly inert.

Below the seal, the sound stopped. The subsonic pressure stopped. The hydrofluoric vapor columns collapsed. Scorivex's projection above them contracted sharply, screaming its acid shriek, and then was pulled back below the sealed vault like smoke being drawn back into a closed bottle.

The Ashveil Plateau was silent.

Fluere sat down directly on the vault seal — not elegantly, but completely, her legs simply no longer interested in holding her. Her wings drooped. The cold aura around her faded to nothing and the temperature of the plateau returned to normal. She looked, for a moment, like exactly what she was: a small, extraordinarily tired chartreuse unicorn sitting on a very old stone.

Caspin sat down beside her immediately, shoulder to shoulder, not saying anything. Sometimes the best chemistry was simple proximity.

Chapter Ten: No Metal, No Magic — And No Better Company

Three weeks after the recasting at Ashveil, Luminance City held what its citizens called a Recognition Gathering — a ceremony rooted in the city's oldest traditions, predating even the Alchemical Wars, in which those who had served the safety of MarBryn and its people were formally acknowledged before the community. It was not a celebration of heroism, exactly. The old tradition held that heroism was simply good citizenship expressed at a larger scale, and that the appropriate response from a civilization was honest gratitude rather than spectacle.

Even so, it was a beautiful morning for it.

The Luminance City central plaza was built above the confluence of the three mineral rivers, on a wide platform of quartz-composite that allowed the water's natural luminescence — a bioluminescent mineral quality of the Veloris tributaries — to glow up through the translucent floor in shifting blues and greens. The moon was visible in the morning sky, an unusual appearance that the city's residents took as a good omen. The plaza was full: scholars, Driftstead traders, engineers, families, alchemical researchers, and ordinary citizens of a world that had come very close to a catastrophe most of them had not known was coming.

Valeria stood at the plaza's edge with her left arm in a chemical-regenerative cast — Sia's formulation, a mineral-polymer matrix infused with healing catalysts that was rebuilding the acid-damaged tissue with remarkable speed. The scar would likely remain. Valeria had examined it in a mirror that morning and decided she did not mind.

She was standing beside Gordas, who was formally reinstalled as Arch-Chancellor of the Chemarcane Academy, a fact that still seemed to slightly surprise him every time it was mentioned. He wore the formal robes with the specific awkwardness of someone who had spent eleven years deliberately not wearing them, but his posture was steady and his expression carried a quiet weight of purpose that seemed appropriate to the role.

"How does it feel?" Valeria asked him.

"Like something that should have happened eleven years ago," he said. "And also like a great deal of work." A pause. "Both things can be true."

"What happens to Sarric?"

"Academy tribunal followed by MarBryn civil judiciary. He will spend a long time explaining himself to a great many people." Gordas watched the plaza fill around them. "He was not wrong that Scorivex is an extraordinary intelligence. He was catastrophically wrong about everything else."

Sia was near the plaza's chemical-display installation — a living exhibit of elemental chemistry she had apparently organized in the two weeks since Ashveil, working with Academy junior faculty who clearly adored her. Mineral samples from across MarBryn were arranged in a luminous display modeled on the periodic table, each element represented by a sample from its natural deposit location: iron-red ores from the Orewall Peaks, copper-green malachite from the plateau flanks, noble gas crystals from the deep Academy vaults. The display drew children to it with the reliable pull of something both beautiful and true.

"No Metal, No Magic," a small girl in the crowd read from the display's heading inscription, sounding the words out carefully.

"That is right," Sia said, crouching to the child's level with genuine delight. "Do you know what it means?"

"That metals make magic happen?"

"It means that everything in the world — every building, every road, every piece of technology, every act of elemental power — is made of the elements. The metals, the gases, the compounds, the chemistry. Without them, none of it exists. The philosophy reminds us to understand what the world is made of and to respect it." Sia touched the iron-ore sample. "And never to use it carelessly. Because chemistry can build everything." A beat. "And it can also dissolve it."

Caspin found Fluere at the far edge of the plaza, where the quartz-composite floor ended and a low railing overlooked the river confluence directly. The water below glowed in slow, shifting patterns of blue-green luminescence. Fluere was resting her forehooves on the railing in the particular way that meant she had been standing there for a while, thinking. Her mane caught the moon morning light and turned each lime and emerald ringlet to something precious. The teal ornament winked quietly near her cheek.

"You look better," Caspin said, coming to stand beside her.

"I feel better," she agreed. She had slept for thirty-six hours after the recasting — a recovery that Sia described as medically appropriate and Caspin described as impressive even by her standards. "The matrix is stable. Sia's morning readings confirm it. The noble-gas suppression envelope is fully restored and she estimates it is stronger than the original, because the fluoride-xenon bonding we introduced is more resistant to acid attack than the pure noble-gas matrix was."

"So we fixed it and made it harder to break," Caspin said.

"Scorivex is contained and will remain so," Fluere said. "Sarric is in custody. The Academy is in good hands." She paused. "For now. MarBryn will face other challenges. The world is always in the process of becoming something, and that process is not always careful."

"Philosophy from the unicorn," Caspin said, smiling.

"I have had a great deal of time to think recently. Some of it while napping." She turned those large, warm dark eyes on him, and they held the sparkle they always carried — that quality of wonder that no amount of danger or exhaustion had ever quite extinguished. "Caspin. Your mother's research on Nullvault containment breach scenarios. The Academy has it now?"

He straightened slightly. "Gordas offered to have it formally included in the Academy's archive. Properly credited." His voice held something that was carefully not quite emotion. "She would have liked that."

"She would have loved it," Fluere said, with the quiet certainty of someone who understood what mattered to people even when they could not say it. "And she would have been proud of you."

Caspin nodded, once, and looked at the glowing water.

The Recognition Gathering began with a formal address from Gordas — concise, serious, historically grounded, and appropriately focused on chemistry rather than drama, which drew appreciative murmurs from the Academy faction of the crowd. Valeria was recognized for her sustained defense above the vault seal, the injury to her arm mentioned with the direct honesty that the tradition required. Sia was recognized for the formulae reconstruction. Caspin for his engineering contributions in the field, with a secondary note that his mother's theoretical work had been integral to the recasting's viability — at which Caspin looked at the quartz floor very carefully for a moment.

When Fluere was recognized, Gordas simply said: "Without the Fluorine Elemental, the matrix recasting was not possible. This world owes her the gratitude that is very difficult to express because it encompasses everything — all the lives that continued because the vault held, all the civilization that persists because the elements were understood and respected and used with care. Fluere of Verdanmoor: "No Metal, No Magic. And today, no magic without you."

The plaza was quiet for a respectful moment before the gathered citizens responded in the traditional way — striking their hands together in the rhythm that meant acknowledged and remembered.

Fluere stood in the copper-tinged morning light on the luminescent quartz platform, her butterfly wings half-spread to balance on her compact hooves, her lime colored meta horn catching the moon's light, and she felt — not proud exactly, though there was something of that — but settled. The way a chemical reaction feels when it reaches its equilibrium. Complete.

Later, as the formal proceedings concluded and the plaza shifted to the community gathering that always followed — mineral-bread, river-caught food, chemistry-light displays, children running on the glowing floor — the five of them drifted together near Sia's elemental display. Five people and one unicorn, standing among their world's periodic table, each element labeled and luminous.

Valeria found Element 9 — Fluorine — and read the display note aloud. "Fluorine: the most reactive of all elements. The most electronegative. It will combine with nearly any substance and drive powerful, lasting bonds. In the right hands, it is one of the most constructive forces in chemistry." She looked at Fluere. "Accurate."

"All descriptions of fluorine are accurate," Fluere said, with mild dignity. "It is a very well-understood element."

"Modest, too," Caspin added.

"Chemistry does not require modesty," Fluere said. "It requires precision." But she was smiling — that warm, unhurried smile that had been there since before any of them knew her, the one that said the world was still full of good things and she intended to find them.

Gordas raised a small glass of mineral-water in the traditional Luminance City fashion, and they all followed suit — Caspin with his engineer's flask, Sia with a sample beaker she had not quite put down all morning, Valeria with the ease of someone growing comfortable in good company.

"No Metal," Gordas said.

"No Magic," the rest of them answered.

And in the bioluminescent light of three rivers meeting beneath a city built from the bones of the elements, that was exactly and entirely enough.

Fluorine Gas

Element Facts

- Isolated in 1886 in France
- Diatomic gas
- N/A to electricity
- Not magnetic
- Halogen

Author's Note

Fluere The Magical Metal Horn Unicorn With Fluorine Magic is my invitation to walk with chemistry as if it were a road laid out by magic itself. This book began as a stubborn question I kept returning to: what if the periodic table were not a dry catalog of facts but a living universe that characters could inhabit, manipulate, and defend with courage, friendship, and wit? The world of MarBryn is my answer to that question—a place where ancient mysticism and modern science share a single purpose: to understand, respect, and responsibly wield the elements that form everything around us.

My aim was to craft a story that pulses with heart-pounding action while offering genuine, accessible threads of science. I wanted Fluere to feel like someone you would want beside you in a crisis: courageous, compassionate, and quietly capable. Her fluorine-based powers—molecular manipulation, cryomancy, and the ability to sense and shape matter—are not just cool tricks; they symbolize the idea that understanding chemistry can empower us to protect what we care about most. The companion characters—Caspin, Valeria, Gordas, Sia, and the others—are designed to be more than plot devices. They represent different facets of curiosity: invention, skepticism, mentorship, courage, and collaboration. Their talents, flaws, and growth arcs mirror the collaborative, sometimes messy, reality of real-world scientific progress.

I also wanted to honor MarBryn's philosophy of No Metal, No Magic in a way that feels respectful rather than prescriptive. This book isn't about choosing science over wonder or vice versa; it's about seeing how they illuminate each other. When Fluere channels her element, and when the team deciphers ancient containment formulae, you're watching a dialogue between past wisdom and present ingenuity—an echo of real-world scientific history where old knowledge informs new discoveries.

If you're reading this, I hope Fluere sparks curiosity about the elements and how they shape the world. And I hope you're reminded that science can be adventurous, collaborative, and full of wonder. May MarBryn inspire you to explore, question, and imagine—with kindness, integrity, and a dash of magic.

Sybrina Durant

About The Author

Sybrina Durant is the author of

"The Magical Elements of the Periodic Table Series"
"The No Metal No Magic Series"
"The Magical Elementals Chemistry Fantasy Series"
"The Blue Unicorn – Journey ToOsm" books

and she is the creator of the "Elemental Fantasy Periodic Table" which is available as a brilliantly colored poster.

There are also book activities, songs, trading cards and t-shirts for all 118 elements and elementals, too. Visit MagicalPTElements.com to see everything that is currently available.